GREEN MACHINES

ECO-BOATS

by Golriz Golkar

Consultant: Beth Gambro
Reading Specialist, Yorkville, Illinois

BEARPORT PUBLISHING

Minneapolis, Minnesota

Teaching Tips

Before Reading

- Look at the cover of the book. Discuss the picture and the title.
- Ask readers to brainstorm a list of what they already know about boats. What can they expect to see in this book?
- Go on a picture walk, looking through the pictures to discuss vocabulary and make predictions about the text.

During Reading

- Read for purpose. Encourage readers to think about what makes a boat an eco-boat as they are reading.
- Ask readers to look for the details of the book. What are they learning about things that can make boats better for Earth?
- If readers encounter an unknown word, ask them to look at the sounds in the word. Then, ask them to look at the rest of the page. Are there any clues to help them understand?

After Reading

- Encourage readers to pick a buddy and reread the book together.
- Ask readers to name something that can make boats eco-friendly. Find the page that tells about this thing.
- Ask readers to write or draw something they learned about eco-boats.

Credits: Cover and title page, © picturelibrary/Alamy; 3, © Cico/Shutterstock and © Sergiy1975/Shutterstock; 5, © Aerial-motion/Shutterstock; 6–7, © Birdiegal/Shutterstock; 9, © Scharfsinn/Alamy; 10–11, © Eric TRAVERS/Getty Images; 13, © The Art of Pics/Shutterstock; 14–15, © Kapitonov Ilia/Shutterstock; 17, © Andreas G. Karelias/Shutterstock; 18–19, © Eskimo/Alamy; 20–21, © YanLev/Shutterstock; 22, © Adrie Oosterwijk/Shutterstock; 23TL, © Goran Jakus Photography/iStock; 23TR, © freevideophotoagency/Shutterstock; 23BL, © Sergey Ryzhov/Shutterstock; and 23BR, © hektoR/Shutterstock.

Library of Congress Cataloging-in-Publication Data

Names: Golkar, Golriz, author.
Title: Eco-boats / by Golriz Golkar.
Description: Bearcub books. | Minneapolis, Minnesota : Bearport Publishing
Company, [2023] | Series: Green machines | Includes bibliographical
references and index.
Identifiers: LCCN 2022002330 (print) | LCCN 2022002331 (ebook) | ISBN
9781636917436 (library binding) | ISBN 9781636917504 (paperback) | ISBN
9781636917573 (ebook)
Subjects: LCSH: Boats and boating--Environmental aspects--Juvenile
literature. | Electric boats--Juvenile literature. | Boats and
boating--Technological innovations--Juvenile literature. | CYAC: Boats
and boating. | LCGFT: Picture books. | Instructional and educational
works.
Classification: LCC VM150 .G58 2023 (print) | LCC VM150 (ebook) | DDC
623.82--dc23/eng/20220203
LC record available at https://lccn.loc.gov/2022002330
LC ebook record available at https://lccn.loc.gov/2022002331

For more information, write to Bearport Publishing, 5357 Penn Avenue South, Minneapolis, MN 55419. Printed in the United States of America.

Contents

Splash!

Take a trip out to sea.

A special boat will bring you there.

This boat is better for ocean plants and animals.

It is an eco-boat!

Some boats use **fuel** to go.

This makes bad gases.

The gases go into the air and water.

They make Earth dirty.

CAP PASLEY
MAERSK
MAERSK SEALAND
CMA CGM

But eco-boats are better.

They may use special fuels that do not make bad gases.

Then, the air and water stay cleaner.

Hydrogen
Hydrogen
1.840
H_2
H_2
P

Some eco-boats get **energy** in another way.

They use **electricity**.

Using electricity does not make gases.

Say electricity like i-lek-TRIS-i-tee

Electricity can come from many things.

Moving water makes electricity for some boats.

The sun's light can make it, too.

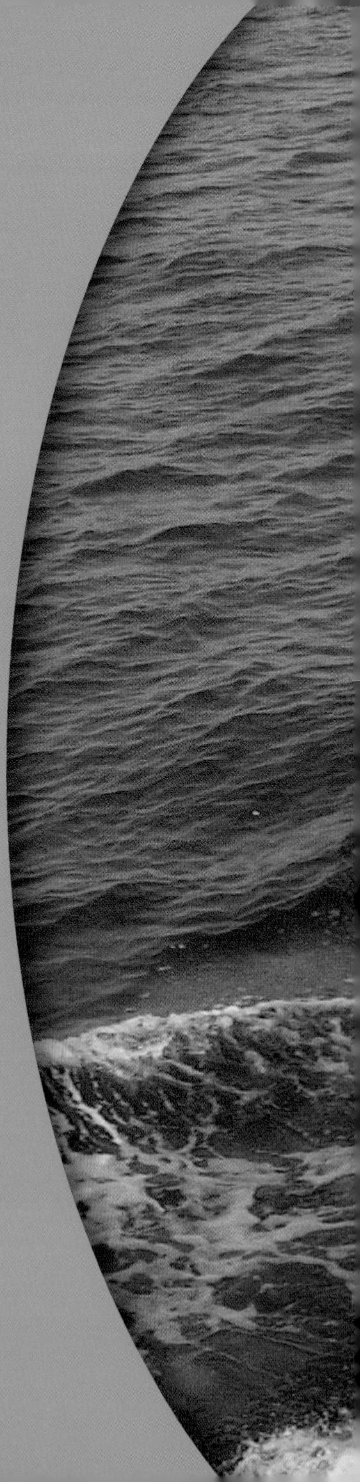

GREENBOAT

Even better, other boats do not use fuel or electricity.

Many sailboats use the wind to go.

Woosh!

La Renaissance

What else can make a boat better for Earth?

Being made differently can!

Some boat paints have **poison** in them.

They make plants and animals sick.

HYBRIDE

Eco-boats use paint that does not have poison.

This keeps the water cleaner.

Plants and animals are safer.

Eco-boats are here to help.

They keep the water safer for all living things.

Let's go green with eco-boats!

Green Boats

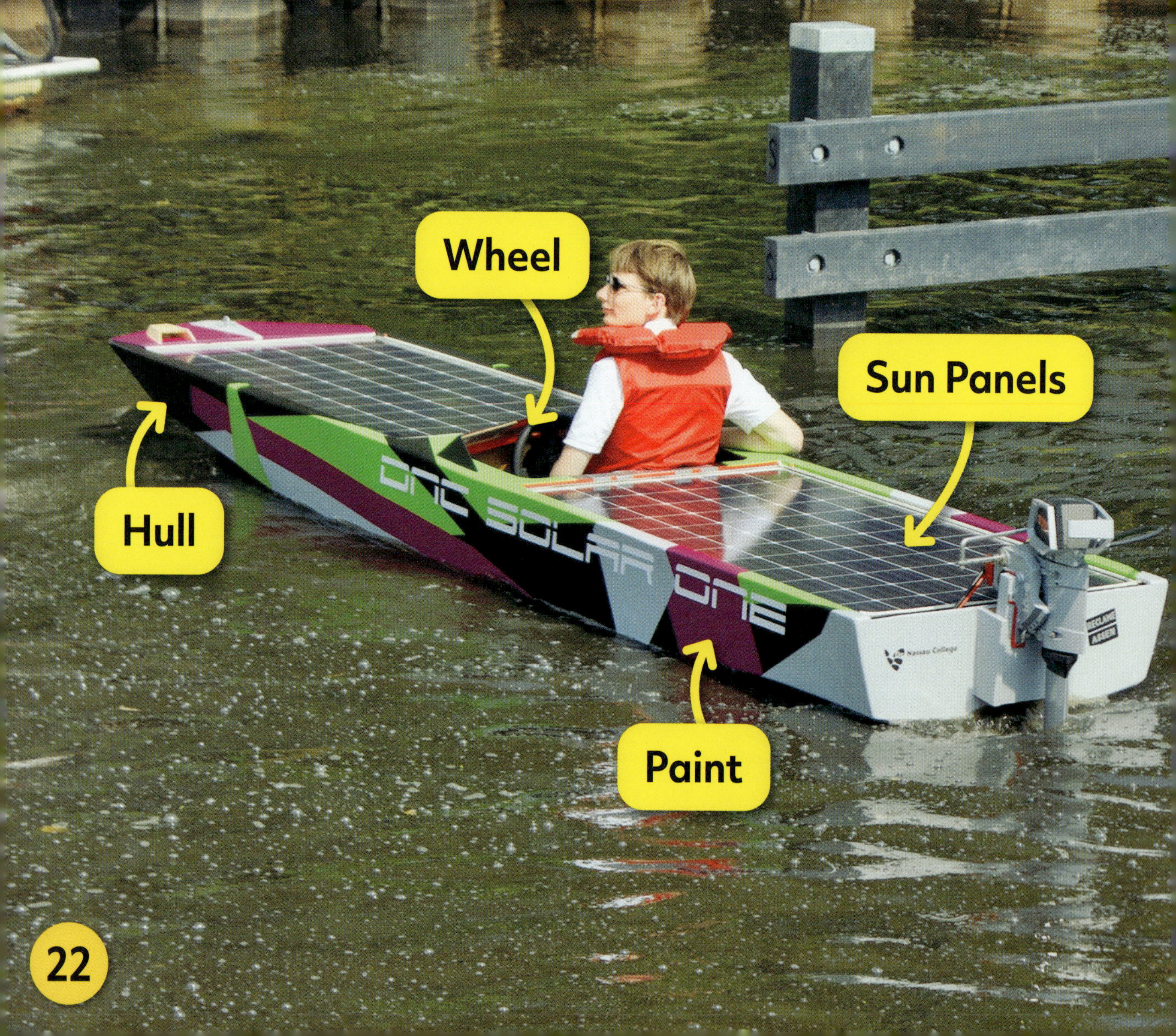

Glossary

electricity a form of energy that can make things work

energy the power machines need in order to work

fuel a liquid used to make some boats go

poison something that can hurt or kill a living thing

Index

Read More

Kenan, Tessa. *Boats (Let's Go!).* Minneapolis: Jump!, Inc., 2019.

Ventura, Marne. *Building Boats That Float (Fun STEM Challenges).* North Mankato, MN: Pebble Books, 2020.

Learn More Online

1. Go to **www.factsurfer.com** or scan the QR code below.
2. Enter "**Eco-Boats**" into the search box.
3. Click on the cover of this book to see a list of websites.

About the Author

Golriz Golkar took her first boat ride when she was six. She saw many whales and dolphins.